Math Challenge the Singapore Way

Grade 2

Copyright © 2012 Marshall Cavendish Corporation

Published by Marshall Cavendish Education

Marshall Cavendish Corp.
99 White Plains Road
Tarrytown, NY 10591
Website: www.marshallcavendish.us/edu

Originally published as Skills in Problem-Solving Maths Copyright © 2002 Times Media Private Limited, Copyright © 2003, 2010 Marshall Cavendish International (Singapore) Private Limited

All rights reserved.

No part of this publication may be reproduced, stored in a retrieval system or transmitted, in any form or by any means, electronic, mechanical, photocopying, recording, or otherwise, without the prior permission of the copyright owner. Request for permission should be addressed to the Publisher, Marshall Cavendish Corporation, 99 White Plains Road, Tarrytown, NY 10591. Tel: (914) 332-8888, fax: (914) 332-1888.

Other Marshall Cavendish Offices:
Marshall Cavendish International (Asia) Private Limited, 1 New Industrial Road, Singapore 536196 • Marshall Cavendish International (Thailand) Co Ltd. 253 Asoke, 12th Flr, Sukhumvit 21 Road, Klongtoey Nua, Wattana, Bangkok 10110, Thailand • Marshall Cavendish (Malaysia) Sdn Bhd, Times Subang, Lot 46, Subang Hi-Tech Industrial Park, Batu Tiga, 40000 Shah Alam, Selangor Darul Ehsan, Malaysia.

Marshall Cavendish is a trademark of Times Publishing Limited

ISBN 978-0-7614-8028-0

Printed in the United States
135642

Introduction

Math the Singapore Way is a term coined to refer to the textbook series used in Singapore schools. Math the Singapore Way focuses on problem solving, given that it is based on a curriculum framework that has mathematical problem solving as its focus. Math the Singapore Way also focuses on thinking, given that the Singapore education system is driven by the *Thinking Schools, Learning Nation* philosophy. Math the Singapore Way is also based on learning theories that provide clear directions on how mathematics is learned and should be taught.

Singapore mathematics textbooks, initial teacher preparation, and the subsequent professional development for teachers are based on helping teachers understand what to teach in mathematics and how to teach it.

While these books are not part of the formal classroom program, they provide selected groups of students with a necessary challenge. Some students complete basic materials easily and enjoy moving on to more challenging tasks. These books are written with that goal in mind. Such learning materials, if they are prepared consistently with the fundamentals of Math the Singapore Way, place an emphasis on problem solving and provide students with various strategies to solve problems, including the use of visuals. It should be noted that good practice is not a matter merely of random repetition. Learners must be challenged through sets of carefully crafted problems. Good practice consists of careful variations in the tasks learners are given.

I hope this series of books is able to provide the necessary help for learners who need to be challenged beyond basic concepts and skills.

Yeap Ban Har
Marshall Cavendish Institute

Preface

Math Challenge the Singapore Way was written to provide students with the practice needed to excel in Math. The problems have been designed in accordance with the latest Math syllabus used in Singapore.

Problems have been differentiated into two levels in each exercise. These exercises test students on their understanding of the mathematical concepts. Questions that allow calculator use are indicated with a 🖩. The more difficult questions are highlighted with a Math Buddy icon .

Worked examples are also provided at the start of each exercise to help students understand the mathematical processes involved.

With this book, students will be exposed to a wide variety of problems that will help them understand math and take their school and state exams with greater confidence.

Notes pages are provided at the back of the book for those who require more space to work out the solutions to the problems.

Contents

Exercise 1   Addition and Subtraction within 100 1

Exercise 2   Addition and Subtraction within 1,000 10

Exercise 3   Multiplication 19

Exercise 4   Division 28

Exercise 5   Length 37

Exercise 6   Weight 47

Exercise 7   Money 56

Exercise 8   Fractions 64

Exercise 9   Time 71

Exercise 10   Volume 80

Exercise 11   Graphs 90

Exercise 12   Lines and Surfaces 97

Exercise 13   Shapes and Patterns 100

Test Yourself 1 103

Test Yourself 2 111

Test Yourself 3 119

Answers 127

Exercise 1

Addition and Subtraction within 100

Example 1

53 tickets were sold on Saturday.
38 tickets were sold on Sunday.
How many tickets were sold altogether?

53 + 38 = 91

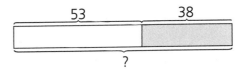

91 tickets were sold altogether.

Example 2

Mary has 25 blue balloons.
She has 14 more red balloons than blue balloons.
How many balloons does Mary have altogether?

25 + 14 = 39

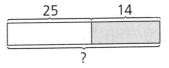

Mary has 39 red balloons.

25 + 39 = 64

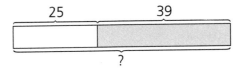

Mary has 64 balloons altogether.

Level 1 Solve the following problems.

1. Wendy had 31 beads.
   Josephine had 38 beads.
   How many beads did they have altogether?

   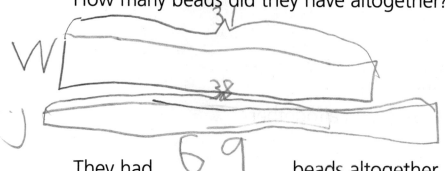

   They had __69__ beads altogether.

---

2. Ms. Ross made 95 cookies.
   She sold 75 of them.
   How many cookies did she have left?

   She had __20__ cookies left.

---

3. A farmer has 47 chicks and 36 ducklings.
   How many more chicks than ducklings does the farmer have?

   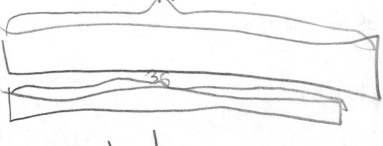

   The farmer has __11__ more chicks than ducklings.

4. A baker made 46 hamburger buns in the morning.
   He made another 41 hamburger buns in the afternoon.
   How many hamburger buns did he make in all?

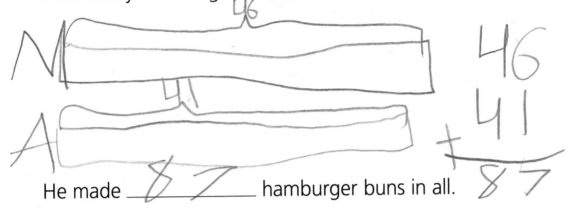

   He made ___87___ hamburger buns in all.

5. 95 men and 64 women work in a bread factory.
   How many more men than women work in the factory?

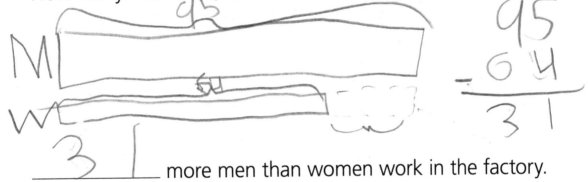

   ___31___ more men than women work in the factory.

6. A green grocer has 83 pineapples.
   He sells 50 of them.
   How many pineapples does he have left?

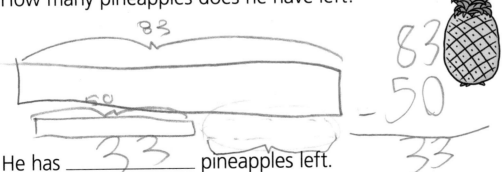

   He has ___33___ pineapples left.

7. There were 64 passengers on a bus.
   14 passengers got off the bus.
   How many passengers were left on the bus?

_____ passengers were left on the bus.

8. There are 28 daisies and 40 roses in a shop.
   How many flowers are in the shop altogether?

There are _____ flowers in the shop altogether.

Level 2 Solve the following problems.

1. A chicken breeder bought 50 eggs.
   After using some eggs, she had 23 eggs left.
   How many eggs did she use?

2. Alice used 38 blue beads, 29 purple beads, and 11 green beads to make a purse.
   How many beads did she use in all?

3. Renee has 33 paper clips.
   She has 17 fewer paper clips than Pamela.
   How many paper clips do both girls have altogether?

4. David had 35 toy cars.
   He gave 9 toy cars to his friends.
   David then bought 25 more toy cars.
   How many toy cars does David have now?

5. Mr. Frank sold 48 digital clocks and had 28 of them left in the showcase.
   How many digital clocks were in the showcase at first?

6. Mr. Adams sold 36 oranges on Sunday, 15 oranges on Monday, and 17 oranges on Tuesday.
   How many oranges did he sell in all?

7. Carol has 27 mangoes and Nora has 60 mangoes.
   How many more mangoes does Carol need in order to have the same number of mangoes as Nora?

---

8. Roger gave 19 pencils to Julio and 24 pencils to Mike.
   Roger then had 35 pencils left.
   How many pencils did Roger have at first?

9. Lily has 47 ribbons.
   She has 8 more ribbons than Jane.
   If Jane has 12 fewer ribbons than Betty, how many ribbons does Betty have?

---

10. A store owner had 61 umbrellas.
    He sold 38 umbrellas and gave away 7 of them as part of a promotion.
    How many umbrellas does he have left?

11. Martin and Lawrence have 92 stickers altogether.
    If Lawrence has 27 stickers, how many more stickers does Martin have than Lawrence?

12. Wilson has 78 balloons.
    Mellie has 52 balloons.
    How many balloons must Wilson give to Mellie so that each of them will have the same number of balloons?

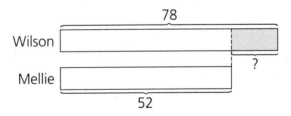

Exercise 2

Addition and Subtraction within 1,000

Example 1

A grocer sold 350 oranges on Monday.
He sold 180 more oranges on Tuesday than on Monday.
How many oranges did he sell altogether?

350 + 180 = 530

The grocer sold 530 oranges on Tuesday.

350 + 530 = 880

The grocer sold 880 oranges altogether.

Example 2

There are 684 students in a school.
297 of them are boys.
How many more girls than boys are in the school?

684 − 297 = 387

There are 387 girls.

387 − 297 = 90

There are 90 more girls than boys in the school.

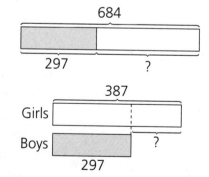

Level 1 Solve the following problems.

1. There are 317 red apples and 182 green apples in a supermarket.
   How many apples are in the supermarket in all?

   There are _____ apples in the supermarket in all.

2. A sales clerk sold 207 handbags in November.
   She sold 281 handbags in December.
   How many handbags did she sell altogether?

   She sold _____ handbags altogether.

3. Donald has 686 marbles.
   Edward has 459 marbles.
   How many more marbles does Donald have than Edward?

   Donald has _____ more marbles than Edward.

4. Jacob has 107 more blue balloons than red balloons.
   If he has 255 blue balloons, how many red balloons does he have?

   He has _____ red balloons.

5. Mary bought 564 beads.
   She used 386 of them to make a necklace.
   How many beads did she have left?

   She had _____ beads left.

6. Janice made 362 paper flowers.
   Donna made 229 paper flowers.
   How many paper flowers did they make altogether?

   They made _____ paper flowers altogether.

7. This table shows the number of visitors at an exhibition on Saturday and Sunday.

Saturday	783
Sunday	957

   How many more visitors went to the exhibition on Sunday than on Saturday?

   _____ more visitors went to the exhibition on Sunday than on Saturday.

8. This chart shows the number of tomatoes and strawberries sold by Mr. Gray.
   How many pieces of fruit did he sell altogether?

   He sold_____ pieces of fruit altogether.

Level 2 Solve the following problems.

1. There are 596 books on one shelf and 348 books on another shelf.
   How many books are on both shelves?

2. Doreen has 117 badges.
   Eliza has 93 more badges than Doreen.
   How many badges does Eliza have?

3. Grace had 215 stamps.
   Margie had 128 stamps fewer than Grace.
   How many stamps did Margie have?

4. A farmer had 767 seeds.
   He planted 268 seeds and gave 145 of them to his neighbor.
   How many seeds does he have left?

5. A grocer had 671 oranges.
   After selling some of them, he had 374 oranges left.
   How many oranges did he sell?

6. Each bag contains some potatoes.
   Wes puts them all in one big box.
   How many potatoes are in the box?

7. There are 435 boys in a school.
   There are 57 more girls than boys in the school.
   How many students are in the school?

8. A farmer had 851 eggs.
   Some of them were broken, and 764 eggs were left.
   How many eggs were broken?

9. There are 145 more cars than motorcycles in a parking garage.
   There are 94 more motorcycles than vans in the parking garage.
   If there are 105 vans, how many cars are there?

10. Greta has 183 more stickers than Shannon.
    Shannon has 499 stickers.
    How many stickers does Greta have?

11. Julia and Paola had 64 beads.
    Paola had 46 fewer beads than Julie.
    How many beads did Paola have?

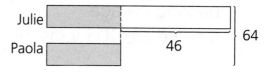

12. Ricardo bought 350 cherries.
    After eating 179 of them, he bought another 246 cherries.
    How many cherries does he have now?

Exercise 3

Multiplication

Example 1

There are 7 students.
Each student makes 3 bows.
How many bows do they make altogether?

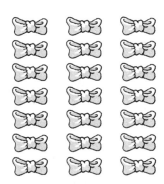

$7 \times 3 = 21$

They make 21 bows altogether.

Example 2

One girl has 4 flowers.
How many flowers do 6 girls have?

$6 \times 4 = 24$

6 girls have 24 flowers.

Level 1 Solve the following problems.

1. There are 2 apples in one bag.
   How many apples are in 4 bags?

There are _____ apples in 4 bags.

2. There are 3 boxes in one row.
   How many boxes are in 3 rows?

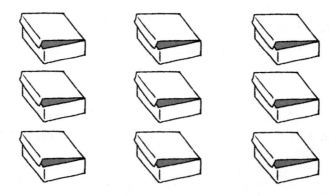

There are _____ boxes in 3 rows.

3. One boy has 6 pens.
   How many pens do 2 boys have?

   2 boys have _____ pens.

4. A table has 4 legs.
   How many legs do 7 tables have?

   7 tables have _____ legs.

5. 1 girl has 6 cookies.
   How many cookies do 5 girls have?

   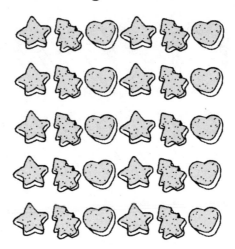

   5 girls have _____ cookies.

6. There are 9 pencils in one bundle.
   How many pencils are in 2 bundles?

   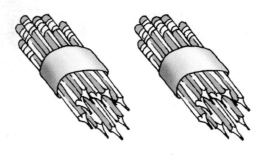

   There are _____ pencils in 2 bundles.

7. Ms. Shore sews 5 blouses a day.
   How many blouses can she sew in 5 days?

   She can sew _____ blouses in 5 days.

---

8. Linda put 10 pears in each of 4 boxes.
   How many pears did she put in 4 boxes?

   Linda put _____ pears in 4 boxes.

Level 2 Solve the following problems.

1. There are 3 boys.
   Each boy has 10 ping pong balls.
   How many ping pong balls do the 3 boys have altogether?

   30

---

2. There were 4 boxes.
   There were 5 watches in each box.
   How many watches were there altogether?

---

3. Ms. Ramirez bought 7 boxes of pencils.
   There were 5 pencils in each box.
   How many pencils did she buy altogether?

4. There are 9 monkeys.
   Each monkey has 3 bananas.
   How many bananas do they have altogether?

   27

5. There are 8 tables in one row.
   How many tables are in 5 rows?

   40

6. 1 box of cookies costs $6.
   How much do 3 boxes of cookies cost?

   $18

7. Daniela buys 3 jars of jelly.
   There are 5 cups of jelly in each jar.
   How many cups of jelly does Daniela buy?

   15

8. Katina puts 9 chicks in each of 4 baskets.
   How many chicks are in 4 baskets?

   36

9. There were 6 girls in 1 group.
   How many girls were in 4 groups?

10. Lisa saves $3 a day.
    How much does she save in 7 days?

    $21

11. There are 10 people at a picnic.
    Each person has 7 bottles of water.
    How many bottles of water do they have altogether?

    70

12. ✪ × ♣ = 24

    ✪ − ♣ = 2

    ✪ = 6
    ♣ = 4

    (a) What is ✪ ?
    (b) What is ♣ ?

Exercise 4

Division

Example 1

24 toy soldiers are shared equally among 4 children. How many toy soldiers does each child get?

$24 \div 4 = 6$

Each child gets 6 toy soldiers.

Example 2

Wendy puts 21 beads equally into 3 boxes. How many beads are in each box?

$21 \div 3 = 7$

There are 7 beads in each box.

Level 1 Solve the following problems.

1. 3 girls share 24 mangoes.
   How many mangoes does each girl get?

   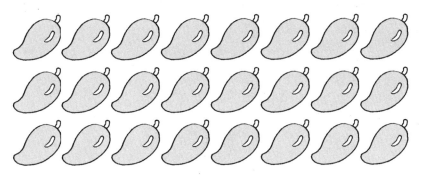

   24 ÷ 3 = ☐

   Each girl gets ____8____ mangoes.

---

2. Melissa puts 12 cupcakes equally on 2 plates.
   How many cupcakes are on each plate?

   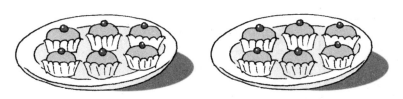

   12 ÷ 2 = ☐

   There are ____6____ cupcakes on each plate.

3. 15 cups of yogurt are put equally into 5 rows.
   How many cups of yogurt are in each row?

   $15 \div \boxed{5} = \boxed{3}$

   There are ____3____ cups of yogurt in each row.

4. Mr. Estevan divided 40 stickers equally among 10 students.
   How many stickers did each student get?

   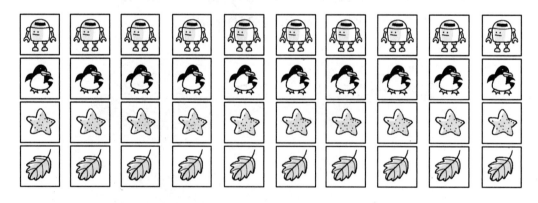

   $40 \div \boxed{10} = \boxed{4}$

   Each student got ____4____ stickers.

5. Bernard puts 27 eggs equally onto 3 trays.
   How many eggs are on each tray?

   $27 \div 3 =$

   There are __9__ eggs on each tray.

---

6. 2 boys shared 14 strawberries equally.
   How many strawberries did each boy get?

   $14 \div 2 =$

   Each boy got __7__ strawberries.

7. 20 flags are arranged equally in 2 rows.
   How many flags are in each row?

   $20 \div 2 =$

   There are ____10____ flags in each row.

---

8. 32 stamps are shared equally among 4 girls.
   How many stamps does each girl get?

   $32 \div 4 =$

   Each girl gets ____8____ stamps.

Level 2 Solve the following problems.

1. Sally and Joan share 18 crayons equally.
   How many crayons does each girl get?

   $18 \div 2 = 9$

---

2. Mr. Dawson made 45 kites.
   5 boys shared the kites equally.
   How many kites did each boy get?

   $45 \div 5 = 9$

---

3. Sean puts 10 candles equally onto 5 cakes.
   How many candles are on each cake?

   2

4. There are 40 lemons.
   Ms. Welles puts them equally into 5 baskets.
   How many lemons are in each basket?

   $40 \div 5 = 8$

---

5. 2 men and 2 women shared 16 peaches equally.
   How many peaches does each person get?

   $16 \div 4 = 4$

---

6. There are 30 fish.
   Terry puts them equally into 5 jars.
   How many fish are in each jar?

   $30 \div 5 = 6$

7. 4 workers share 24 nails equally.
   How many nails does each worker get?

   6

8. Ms. Morgan bought 50 bookmarks.
   She shared them equally among her 5 children.
   How many bookmarks did each child get?

   10

9. David and his 4 friends shared 20 pens equally. How many
   pens did each of them receive?

   4

10. There are 16 boys and 20 girls in a class.
    All the students are grouped into 4 equal teams.
    How many students are on each team?

    36

    9

11. Jessica made 25 hot dog buns.
    She put them equally into 5 boxes.
    How many buns were in each box?

    5

12.

    🦋 = 10

    🐞 = 5

Exercise 5

Length

Example 1

An oak tree is 21 feet tall.
An elm tree is 6 feet shorter than the oak tree.
Find the height of the elm tree.

21 − 6 = 15 feet

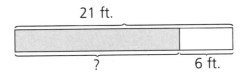

The height of the elm tree is 15 feet.

Example 2

Pedro is 60 inches tall.
Walt is 54 inches tall.
What is their total height?

60 + 54 = 114 in.

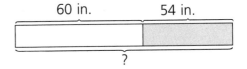

Their total height is 114 inches.

Level 1 Solve the following problems.

1. (a) Draw a line 2 inches shorter than the pen.
   (b) Find the length of the line.

   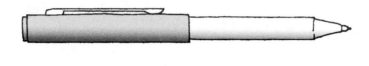

   (b) The length of the line is __1 1/2__ inches.

2. Use your ruler to measure the three lines in ~~centimeters~~ inches.
   (a) What is the total length of the three lines?
   (b) How much longer is Line C than Line B?

   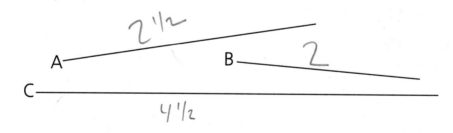

   (a) The total length of the three lines is __9__ ~~centimeters~~ inches.

   (b) Line C is __2 1/2__ inches ~~centimeters~~ longer than Line B.

3. Wendy bought 458 yards of pink ribbon and 379 yards of purple ribbon.
How much ribbon did she buy in all?

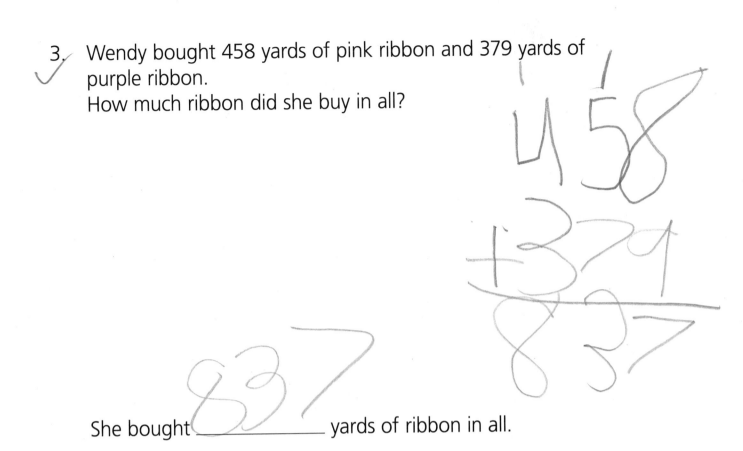

She bought _____837_____ yards of ribbon in all.

4. A piece of wire is bent to form a triangle.
Find the length of the piece of wire.

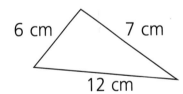

The piece of wire is _____25_____ centimeters long.

5. What is the total length of the three planks?

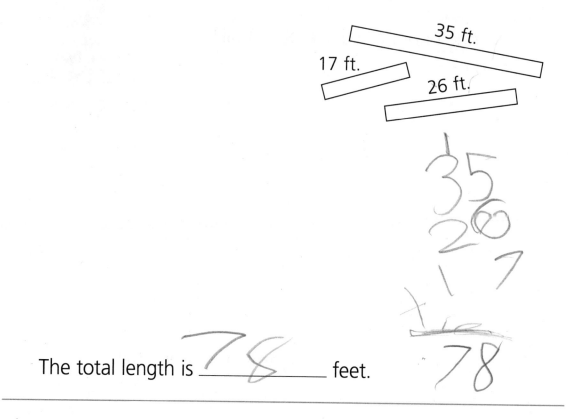

The total length is __78__ feet.

6. A bolt of cloth was 55 yards long.
   The store owner used 39 yards of the cloth.
   How many yards of cloth were left?

__16__ yards of cloth were left.

7. Benjamin ran 90 yards in the morning and another 95 yards in the evening.
How many yards did he run that day?

He ran __185__ yards that day.

8. How far is the boy from his apartment?

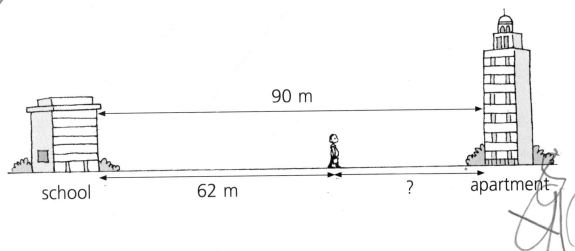

The boy is __28__ meters from his apartment.

Level 2 Solve the following problems.

1. The total length of two pieces of string is 425 yards. If one piece of string is 268 yards long, what is the length of the other piece of string?

2. There are 3 ropes.
   Each rope is 6 feet long.
   Find their total length when they are joined end to end.

3. A piece of ribbon, 12 yards long, is cut into 3 equal pieces. What is the length of each piece?

4. A rope, 580 feet long, is cut into 2 pieces.
   One piece is 193 feet.
   What is the length of the other piece?

---

5. Tammy walked 275 feet to the library and then walked another 180 feet to the bus stop.
   How far did she walk?

---

6. Ken has 32 inches of wire. He cuts it into 4 equal pieces.
   How long is each piece of wire?

7. The length of a swimming pool is 50 yards.
   Jerry swims 2 lengths of the swimming pool.
   How many yards does he swim?

8. Jason walked from his house to the library, then to the supermarket and back home.
   If the total distance he covered was 570 meters, find the distance between the library and the supermarket.

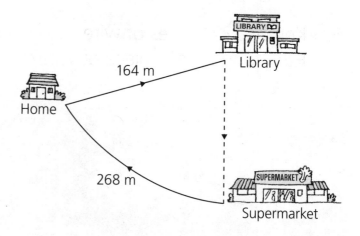

9. Nora has 2 sticks.
   One of them is 46 inches long and the other is 28 inches long.
   She puts them in a row.
   What is the total length of the two sticks?

---

10. Steven cut a sheet of paper into 3 equal pieces.
    Each piece of paper was 8 inches long.
    How long was the sheet of paper before he cut it?

11. A rod is 264 inches long.
    A plank is 178 inches shorter than the rod.
    Find their total length.

    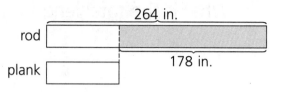

12. Patrick is 154 centimeters tall.
    Wilson is 18 centimeters shorter than Patrick.
    Jerry is 7 centimeters taller than Wilson.
    How tall is Jerry?

    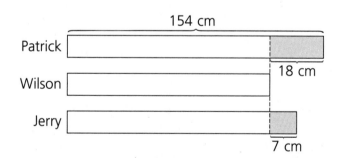

Exercise 6

Weight

Example 1

Ms. Roberts bought 18 pounds of cantaloupes and 29 pounds of bananas. How many pounds of fruit did she buy altogether?

18 + 29 = 47 lb.

She bought 47 pounds of fruit altogether.

Example 2

Hanna weighs 64 pounds
Anita is 7 pounds lighter.
How much does Anita weigh?

64 − 7 = 57 lb.

Anita weighs 57 pounds.

Level 1 Work out the following problems.

1. Linda's weight is 60 pounds. Her sister's weight is 45 pounds.
   How much heavier is Linda?

   Linda is _____ pounds heavier.

2. Doris's weight is 53 pounds. Albert is 18 pounds heavier. How much does Albert weigh?

   Albert weighs _____ pounds.

3. A desk has a weight of 750 ounces.
   A chair has a weight of 250 ounces.
   What is their total weight?

   Their total weight is _____ ounces.

4. Diana bought a table that weighed 556 ounces and a lamp that weighed 329 ounces.
   How much heavier was the table?

   The table was _____ ounces heavier.

5. Mr. Williams, a store owner, had 836 pounds of shrimp.
   He sold 459 pounds of the shrimp.
   How many pounds of shrimp did he have left?

   He had _____ pounds of shrimp left.

6. An empty tank weighs 327 ounces.
   596 ounces of water is poured into the tank.
   Find the total weight of the tank and the water.

   The total weight of the tank and the water is _____ ounces.

7. A potter needs 450 ounces of clay to make some pots.
He has only 265 ounces of clay.
How much more clay does he need?

He needs _____ ounces more clay.

8. Anita weighs 62 pounds She is 6 pounds lighter than Sharon.
What is Sharon's weight?

Sharon weighs _____ pounds.

Level 2 Solve the following problems.

1. 10 books and a bookcase weigh 950 ounces.
   The bookcase weighs 586 ounces.
   What is the weight of the 10 books?

2. Ms. Lewis, a caterer, bought 100 ounces of turkey, 145 ounces of steak, and 275 ounces of ham.
   How many ounces of meat did she buy in all?

3. A horse weighs 875 pounds.
   It is 689 pounds heavier than a deer.
   What is the weight of the deer?

4. Mr. Forest sold 35 pounds of sugar in the morning.
   He sold another 37 pounds of sugar in the afternoon.
   How many pounds of sugar did he sell in all?

---

5. A couch with 4 people on it weighs 923 ounces.
   The couch weighs 347 ounces when no one is on it.
   Find the weight of the four people.

---

6. A fish market had 83 pounds of fish.
   If it sold 46 pounds of fish, how many pounds of fish did it have left?

7. Carol used 390 grams of flour to make a pizza.
   She used 430 grams of flour to make a cake.
   She still had 126 grams of flour left.
   How many grams of flour did she have at first?

   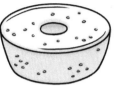

---

8. A computer weighs 315 ounces.
   A printer is 168 ounces lighter than the computer.
   (a) Find the weight of the printer.
   (b) What is the total weight of the computer and the printer?

9. Margaret bought 18 pounds of sugar.
   She put the sugar equally into 3 bags.
   How many pounds of sugar did she put into each bag?

---

10. For the cookout, Alice bought 245 ounces of pumpkin, 180 ounces of squash, and 68 ounces of tomatoes. What was the total weight of the vegetables she bought?

11. Box A weighs 210 grams.
    Box B is 85 grams heavier than Box A and 60 grams lighter than Box C.
    Find the weight of Box C.

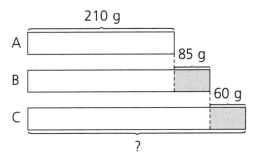

12. The total weight of a crab and a hake fish is 800 grams.
    The fish is 200 grams lighter than the crab.
    What is the weight of the crab?

Exercise 7

Money

Example 1

The pizza costs 70¢ more than the ice cream.
How much does the pizza cost?

$4.35 + 70¢ = $5.05

The pizza costs $5.05.

Example 2

Mike paid $6.25 for a burger and a drink.
The burger cost $4.
How much did the drink cost?

$6.25 − $4 = $2.25

The drink cost $2.25.

Level 1 Solve the following problems.

1. Ms. Harper had $8.60.
   She bought a purse for $3.
   How much money did she have left?

   She had _____ left.

2. Jane has $14 in her wallet.
   Her mother gives her $5.60.
   How much money does Jane have now?

   Jane has _____ now.

3. Ms. Chase bought a bottle of olive oil and a carton of milk.
   How much did she spend altogether?

   She spent _____ altogether.

4. Lillian saved $7.90.
   Elaine saved $6.00.
   How much did they save altogether?

   They saved _____ altogether.

5. Dora spent $11 on Monday and $13 on Tuesday.
   How much did she spend altogether?

   She spent _____ altogether.

6. Kenneth has 95¢.
   Terry has 55¢.
   How much more money does Kenneth have than Terry?

   Kenneth has _____ more than Terry.

---

7. Ms. Yang bought a papaya for $2, a watermelon for $4, and a pineapple for $5.
   How much did she spend in all?

   She spent _____ in all.

Level 2 Solve the following problems.

1. William had $27.80.
   He had $19 left after buying a cake.
   How much did the cake cost?

2. Michael has a 50-dollar bill.
   Which two things can he buy using as much money as possible?

Bag $22.80 Book $24
DVD $35 Pen $13

3. The mangoes weigh 9 kilograms.
   How much do they cost?

---

4. Kerry worked for 4 hours.
   He was paid $10 an hour.
   How much money did he make?

5. Julia wanted to buy a tub of ice cream and a bar of chocolate.
   (a) How much did the two items cost?
   (b) If she had only $12, how much money would she need to be able to pay for the two items?

6. Mr. Brown bought 2 pears, 3 apples, and 1 pineapple. How much did he pay altogether?

2 pears	=
3 apples	=
1 pineapple	=
Total cost	=

7. After paying $15.00 to the cashier, Paul had $25.80 left.
   How much did he have at first?

---

8. Eva bought 3 chicken breasts and a carton of milk for $9.
   The carton of milk cost $3.
   How much did each chicken breast cost?

```
Chicken breast [        ]
Chicken breast [        ]
Chicken breast [        ]  } $9
         Milk  [        ]
                  $3
```

Exercise 8

Fractions

Example 1

Arrange the fractions $\frac{1}{4}$, $\frac{1}{5}$, and $\frac{1}{3}$ in order, beginning with the smallest.

From the model, we can see that $\frac{1}{5}$ is the smallest, followed by $\frac{1}{4}$, and then by $\frac{1}{3}$.

Answer: $\frac{1}{5}$, $\frac{1}{4}$, $\frac{1}{3}$

Example 2

$\frac{3}{4}$ of the members of a youth choir are girls. What fraction of the members are boys?

$$1 - \frac{3}{4} = \frac{4}{4} - \frac{3}{4}$$
$$= \frac{1}{4}$$

$\frac{1}{4}$ of the members are boys.

Level 1 Answer the following questions.

1. What fraction of the circle is shaded?

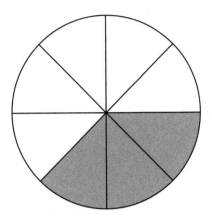

2. What fraction of the rectangle is shaded?

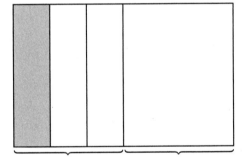

3. Shade $\frac{5}{8}$ of this circle.

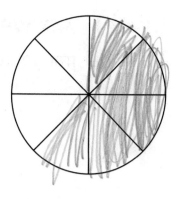

---

4. What fraction of this square is not shaded?

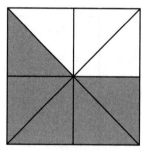

5. Which one of the following is the largest fraction?

   $\frac{1}{4}$, $\frac{1}{3}$, $\frac{1}{5}$, $\frac{1}{6}$

6. Which one of the following is the smallest fraction?

   $\frac{5}{5}$, $\frac{5}{7}$, $\frac{5}{8}$, $\frac{5}{9}$

7. $\boxed{\ }$ + $\frac{3}{8}$ make 1 whole.

   What is the missing fraction in the box?

8. How many fifths are there in 1 whole?

Level 2

1. $\frac{3}{5}$ of the audience in a movie theater are adults and the rest are children.
   What fraction of the audience are children?

2. $\frac{1}{5}$ of the students in a class are boys.
   What fraction of the students in the class are girls?

   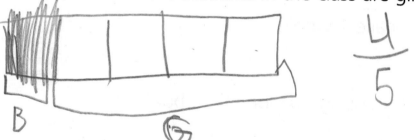

3. Ms. Lee baked a pizza.
   She ate $\frac{1}{8}$ of the pizza and gave $\frac{2}{8}$ of it to her neighbor.
   What fraction of the pizza did she have left?

   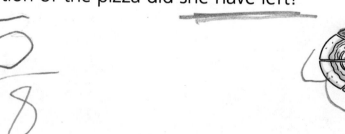

4. A watermelon was cut into 12 equal pieces.
   Maria ate 2 pieces and Ali ate 3 pieces.
   What fraction of the watermelon was left?

   $\frac{7}{12}$

5. A cake was cut into 10 equal pieces.
   Lily ate a few pieces.
   If 7 pieces were left, what fraction of the
   cake had Lily eaten?

   $\frac{3}{10}$

6. Betty cut a cake into equal pieces.
   She ate 2 pieces and gave 5 pieces to her brother.
   If there were 5 pieces left, what fraction of the cake did
   Betty give to her brother?

   $\frac{5}{12}$

   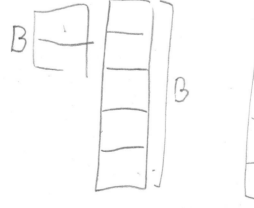

7. Arrange the following fractions in order, from the smallest to the largest.

$$\frac{3}{3}, \quad \frac{3}{4}, \quad \frac{3}{5}, \quad \frac{3}{7}$$

---

8. Eli baked a sponge cake.
   He gave half of the cake to his neighbor.
   He then cut the remaining half into 3 equal pieces and ate 1 piece.
   What fraction of the sponge cake did Eli eat?

Exercise 9

Time

Example 1

The minute hand shows 35 minutes.

Example 2

The time is 4:55.

Example 3

 → 30 min. later →

It is 8 o'clock.   It is 8:30.

Example 4

 → 50 min. later →

It is 2 o'clock.   It is 2:50.

Example 5

 1 hr., 45 min. later

It is 12 o'clock.   It is 1:45.

Example 6

 30 min. later

It is 5 o'clock.   It is 5:30.

Level 1 Complete each of the following:

1.

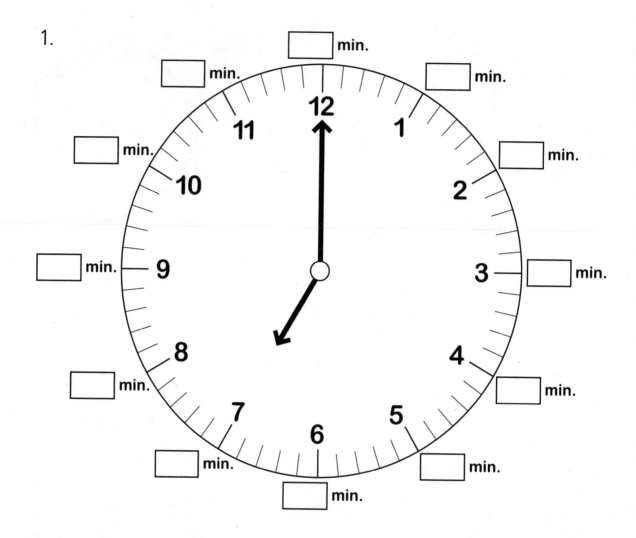

Fill in the blanks.

2.  The minute hand shows _____ minutes.

3.  The minute hand shows _____ minutes.

4.  The minute hand shows _____ minutes.

What time is it?

5.  The time is _____.

6.  The time is _____.

7.  The time is _____.

Level 2 Draw the hands on the clock.

1.

   The time is 8:15.

2.

   The time is 1:55.

3.

   The time is 4:30.

4.

   The time is 3:45.

Fill in the blanks.

5.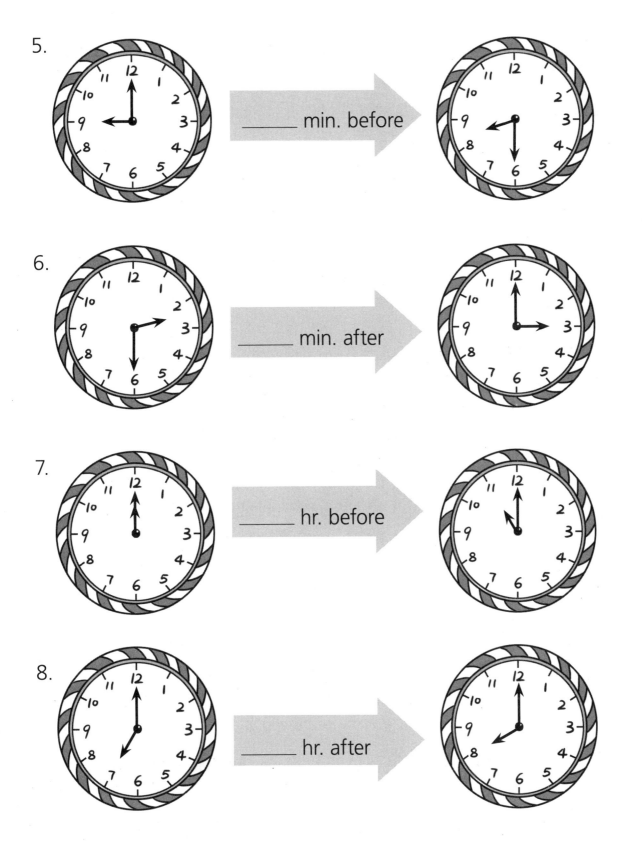

_____ min. before

6. _____ min. after

7. _____ hr. before

8. _____ hr. after

Write a.m. or p.m.

9.

Grandfather goes for a morning walk at 7 _____.

10.

After dinner at 7:30 _____, Marcie helps wash the dishes.

11. Thomas watched a movie.
    It started at 8:30 at night.
    It lasted 1 hour.
    At what time did the movie end?
    Write a.m. or p.m. after the time.

12. Lily started her homework at 9:30.
    She finished her homework at 11:30.
    How long did she take to do her homework?

Exercise 10

Volume

Example 1

A basin contains 28 liters of water.
David pours another 13 liters of water into it.
How many liters of water are in the basin now?

28 + 13 = 41 L

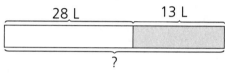

There are 41 liters of water in the basin now.

Example 2

A tank contained 54 gallons of water.
Peggy used 27 gallons of it.
How many gallons of water were left?

54 − 27 = 27 gal.

27 gallons of water were left.

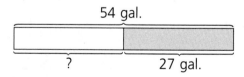

Level 1 Solve the following problems.

1. How many liters of water are in the two containers?

   pail

   pitcher

   There are _____ liters of water in the two containers.

---

2. Mr. Young had 458 gallons of soft drinks.
   He sold 178 gallons of soft drinks.
   How many gallons of soft drinks did he have left?

   He had _____ gallons of soft drinks left.

3. Tank A contains 581 liters of water.
   Tank B contains 239 liters of water.
   How many liters of water are in the two tanks?

   There are _____ liters of water in the two tanks.

4. A pail contains 28 quarts of water.
   A can contains 9 quarts of water.
   How many more quarts of water does the pail contain than the can?

   The pail contains _____ more quarts of water than the can.

5. After using 85 quarts of water, Donald had 247 quarts of water left.
   How many quarts of water did he have at first?

   $$\begin{array}{r} 247 \\ +85 \\ \hline 332 \end{array}$$

   He had __332__ quarts of water at first.

---

6. There are 22 liters of water in a pail and 48 liters of water in a barrel.
   How many liters of water are in the two containers altogether?

   $$\begin{array}{r} 22 \\ +48 \\ \hline 70 \end{array}$$

   There are __70__ liters of water in the two containers altogether.

7. Two basins contain 18 gallons of water altogether.
One of the basins contains 7 gallons of water.
What is the volume of water in the other basin?

The volume of water in the other basin is __11__ gallons.

---

8. Margie bought 3 liters of juice.
Lily bought 2 liters of juice, and Sally bought 6 liters of juice.
How many liters of juice did they buy altogether?

They bought __11__ liters of juice altogether.

Level 2 Solve the following problems.

1. A dairy farmer sold 337 gallons of milk on the first day.
   She sold 497 gallons of milk on the second day.
   How much milk did she sell altogether?

2. There are 608 liters of syrup altogether in two containers.
   There are 259 liters of syrup in one of the containers.
   How many liters of syrup are in the other container?

3. A basin contains 8 quarts of water.
   A tank contains 10 times as much water as the basin.
   How many quarts of water does the tank contain?

4. A tank can hold 265 liters.
   It contains 89 liters of water.
   How much more water is needed to fill the tank?

   _____

5. 45 gallons of milk are shared equally among 5 classes.
   How many gallons of milk does each class get?

   _____

6. Harry bought 7 bottles of apple juice.
   Each bottle contained 2 liters of apple juice.
   How many liters of apple juice did he buy?

7. Mr. Terry had 24 pints of oil.
   He poured them equally into 3 bottles.
   How many pints of oil were in each bottle?

8. How many liters of water will the container hold?

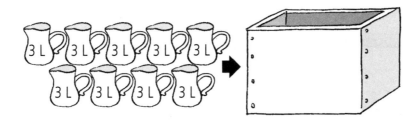

9. A tank contains 64 gallons of water.
   A basin contains 48 gallons less water than the tank.
   A pail contains 8 gallons more water than the basin.
   How many gallons of water are in the pail?

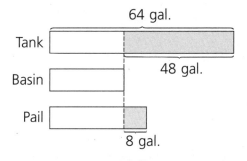

---

10. How many kettles can the pail fill?

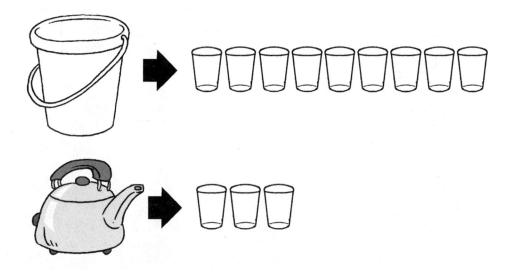

11. Ms. Tomas fills 3 pitchers with juice.
    Pitcher A has twice as much juice as Pitcher B.
    Pitcher B has 4 times as much juice as Pitcher C.
    If Pitcher C has 2 quarts of juice, how much juice is in Pitcher A?

    Pitcher A [|]
    Pitcher B [| | |]
    Pitcher C [ ]
              2 qt.

---

12. A tank has 20 liters of water when it is full.
    There are 15 liters of water in the tank now.
    I use some water from the tank to fill 2 pitchers completely.

    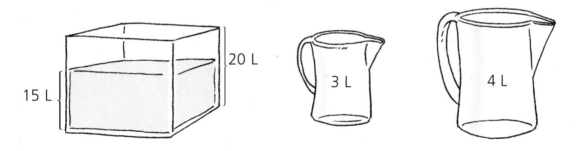

    (a) How much water is left in the tank?
    (b) How much water should be added to the tank to fill it completely?

Exercise 11

Graphs

Example

Study the graph carefully.
Then answer the following questions.

Each Person's Savings	
Martin	△ △ △ △ △
Sally	△ △ △ △
Wesley	△ △
Samantha	△ △ △ △
Each △ stands for $5.	

(a) How much money does Martin have?

Martin has $25.

(b) Which two people have the same amount of money?

Sally and Samantha have the same amount of money.

(c) How much more money does Sally have than Wesley?

$20 − $10 = $10

Sally has $10 more than Wesley.

Level 1   Answer the following questions.

1. Study the graph carefully.
   Then, write the correct answer in each blank.

Number of Strawberries Each Person Has
Mark — 3 strawberries; Stan — 6 strawberries; Teresa — 7 strawberries; Jane — 4 strawberries
Each 🍓 stands for 2 strawberries.

   (a) _____ has the most strawberries.

   (b) _____ has the fewest strawberries.

   (c) Teresa has _____ strawberries more than Jane.

   (d) _____ has 6 strawberries fewer than Stan.

2. Study the graph carefully.
   Then, fill in the blanks with the correct answers.

Games People Play	
Basketball	😊😊😊😊
Soccer	😊😊😊😊😊😊😊
Volleyball	😊😊😊😊😊
Tennis	😊😊😊😊😊
Football	😊😊😊😊😊😊😊😊

   Each 😊 or 😊 stands for 3 people.

   (a) Which is the most popular game? _____

   (b) Which is the least popular game? _____

   (c) How many more people play volleyball than basketball? _____

   (d) What is the total number of people who play soccer and football? _____

Math Challenge the Singapore Way

Level 2 Answer the following questions.

1. Study the graph.
   Then, fill in the blanks with the correct answers.

Number of Bags Sold				
Monday	Tuesday	Wednesday	Thursday	Friday

   Each ▭ stands for 4 bags.

   (a) The largest number of bags were sold on _____.

   (b) _____ fewer bags were sold on Tuesday than on Friday.

   (c) _____ more bags were sold on Wednesday than on Monday.

   (d) 12 more bags were sold on Thursday than on _____.

   (e) _____ bags were sold during the first three days altogether.

2. Study the graph.
   Then, fill in the blanks with the correct answers.

Colors People Like	
Red	◯ ◯ ◯ ◯ ◯
Yellow	◯
Blue	◯ ◯ ◯
Purple	◯ ◯ ◯ ◯
Orange	◯ ◯
Each ◯ stands for 5 people.	

   (a) Which color do people like least?

   _____

   (b) _____ people like orange.

   (c) 5 people like _____.

   (d) _____ people like red and blue.

   (e) How many fewer people like yellow than purple?

   _____

   (f) How many more people like red than orange?

   _____

3. The table below shows the cookies made by a group of boys.

24	12	6	15
Butter cookies	Chocolate chip cookies	Cashew nut cookies	Almond cookies

   (a) Use the information in the table above to make a picture graph.

Butter cookies	
Chocolate chip cookies	
Cashew nut cookies	
Almond cookies	
Each ◯ stands for 3 cookies.	

   (b) _____ more butter cookies were baked than almond cookies.

   (c) The total number of cookies baked was _____.

4. If 🌸 stands for 4 flowers,

   _____ stand for 20 flowers.

---

5. If each ✏️ stands for 5 pencils,

   ✏️ ✏️ ✏️ ✏️ stand for _____ pencils.

---

6. ♥ = ☀ + ☀ + ☺

   ☀ = ☺ + ☺ + ☺

   How many ☺ = ♥ ?

Exercise 12

Lines and Surfaces

Example 1

How many curves are in this figure?

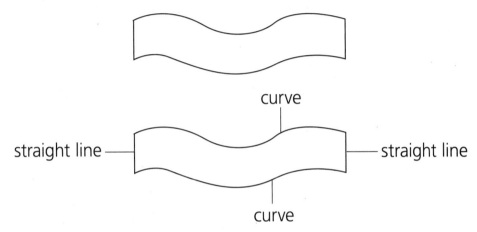

There are 2 curves in this figure.

Example 2

How many flat surfaces are on this object?

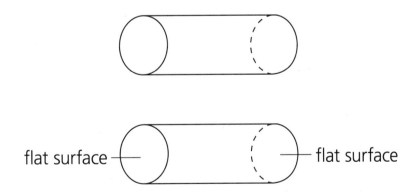

There are 2 flat surfaces on this object.

Level 1 Answer the following questions.

1. How many curves are in the figure?

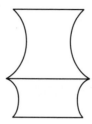

---

2. Which of these letters have both curves and straight lines?

   A   B   C   D   E   F   G

---

3. How many straight lines are in the figure?

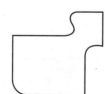

Level 2 Answer the following questions.

1. Find the number of flat surfaces in the following figures.

   (a)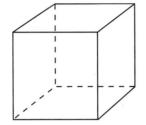

   (b)

   (c)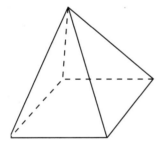

   (d)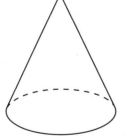

Exercise 13

Shapes and Patterns

Example 1

How many semicircles are in this figure?

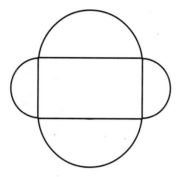

There are 4 semicircles in this figure.

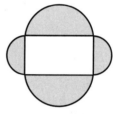

Example 2

Name the shapes that make up this figure.

This figure is made up of a quarter circle, a rectangle, a square, a semicircle, and a triangle.

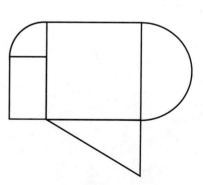

Level 1 Answer the following questions.

1. How many semicircles are in the figure?

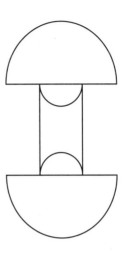

_____

2. Draw an object on the lines to complete the following patterns.

(a) 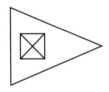 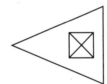 _____

(b)   _____

(c)   _____

Level 2 Answer the following questions.

1. Draw 2 straight lines in the figure below to form 3 triangles.

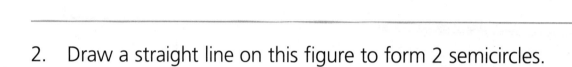

2. Draw a straight line on this figure to form 2 semicircles.

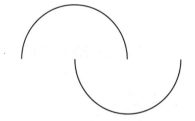

3. Name the following figures.

(a) _____

(b) _____

(c) _____

(d) _____

(e) _____

(f) _____

Test Yourself 1

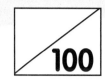

**Do these problems.
Show your work clearly.**

1. A container can hold 36 liters of water.
   Curt pours 4 pails of water to fill it completely.
   How many liters of water does each pail contain?   (5)

2. Letitia gave the cashier $50 for a basket that cost $23.
   How much change would she get from the cashier?   (5)

3. Anthony has 45 stamps.
   If he pastes them equally on 5 post cards, how many stamps does he paste on each postcard?   (5)

4. Mr. Alejo has 2 daughters and 1 son.
   He distributes 24 balloons equally among his children.
   How many balloons does each child get? (5)

---

5. One apple cost 50 cents.
   Ms. Wilson bought 2 apples.
   How much did she spend in all? (5)

---

6. Gail paid $5 for each box of cookies.
   How many boxes of cookies could she buy with $40? (5)

7. 2 pineapples weigh 813 grams.
   One of the pineapples weighs 396 grams.
   What is the weight of the other pineapple? (5)

8. If one piece of rope is 7 feet long, what is the total length of 4 similar pieces of rope? (5)

9. Larry has two 25-cent coins.
   William has five 10-cent coins.
   How much money do they have altogether? (5)

10. A farmer had two cows.
    The first weighed 167 kilograms. The other cow was
    18 kilograms lighter than the first cow.
    Find the total weight of the two cows.   (5)

    ```
                          167 kg
    First cow   [                    ]
                                      18 kg
    Second cow  [                 ]
    ```

---

11. Tracy decorates 9 Christmas trees.
    She uses 2 yards of ribbon for each tree.
    How many yards of ribbon does she use in all?   (5)

12. Ms. Conner bought a pen and a book for $42.
    If the pen cost $17, how much did the book cost?
                                                    (5)

13.

Ricardo	☐ ☐ ☐ ☐ ☐ ☐ ☐
Isabel	☐ ☐ ☐ ☐ ☐
Each ☐ stands for 5 bookmarks.	

The table above shows the number of bookmarks Ricardo and Isabel have.
How many fewer bookmarks does Isabel have than Ricardo?
                                                    (5)

14. A string 28 meters long is cut into 4 equal pieces. Find the length of each piece. (5)

---

15. Tom had $30.
    He bought a container of chicken noodle soup for $4.
    How much money did he have left? (5)

16. 3 hot dog buns are packed into one box.
    How many boxes will be needed to pack 30 hot dog buns?
    (5)

17. There are 10 roses in one bundle.
    A florist has 7 bundles of roses.
    How many roses does the florist have altogether? (5)

18. Kurt had 34 stickers.
    Wilson had 15 fewer stickers than Kurt.
    How many stickers did they have altogether?
    (5)

19. Roland saves $47 more than Matthew.
    Steven saves $16 more than Matthew.
    How much more does Roland save than Steven? (5)

20. A farmer had 394 eggs.
    He sold 189 eggs and broke 23 of them.
    How many eggs did he have left? (5)

Test Yourself 2

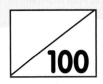

**Do these problems.
Show your work clearly.**

1. There are 8 oranges in one basket.
   How many oranges are in 3 baskets? (5)

2. In a box, there are 25 blue marbles, 18 red marbles, and 27 yellow marbles.
   How many marbles are in the box altogether? (5)

3. Jack has $36.95.
   He needs $18 more to buy a guitar.
   How much does the guitar cost? (5)

4. A basin can hold 9 gallons of water.
   3 basins of water will fill a tank completely.
   How many gallons of water can the tank hold? (5)

---

5. Alejandro gave 24 oranges to his 3 brothers.
   If each brother received an equal number of oranges, how many oranges did each of his brothers receive?
   (5)

---

6. 4 grapefruits cost $12.
   Find the cost of each grapefruit. (5)

7. Package A weighs 524 grams.
   Package B is 135 grams lighter than Package A.
   (a) Find the weight of Package B.
   (b) Find their total weight. (5)

Package A   Package B

8. Sofia bought 5 bottles of cooking oil.
   Each bottle contained 2 quarts of cooking oil.
   How many quarts of cooking oil did Sofia buy altogether? (5)

9. Jessica placed 35 bowls into 5 equal stacks.
   How many bowls were in each stack? (5)

10. 10 students took part in a tennis competition.
    Twice as many students took part in a swimming competition.
    How many students took part in the swimming competition? (5)

11. A stone weighs 3 kilograms.
    A rock is 19 kilograms heavier than the stone.
    (a) Find the weight of the rock.
    (b) Find the total weight of the stone and the rock. (5)

12. A group of people bought a total of 18 sandwiches from a deli.
    Each person bought 3 sandwiches.
    How many people were there? (5)

13. Fernando bicycled from the library to the grocery and then to his home.
    How far did he cycle altogether? (5)

    Library   Grocery   Home
         97 yd.   84 yd.

14. Jane had 570 centimeters of string.
    After using some of it, she had 284 centimeters of string left.
    How many centimeters of string did she use? (5)

---

15. 15 people went to dinner in 3 cars.
    If each car had the same number of people, how many people were in each car? (5)

16. Mr. Miller poured 47 liters of gasoline out of a tank.
    38 liters of gas were left in the tank.
    How much gas was in the tank at first? (5)

---

17. 157 men, 283 women, and 509 children were at a party.
    How many more children than adults were at the party? (5)

    ```
                    157      283
            Adults [    |        ]
          Children [             ]
                         509
    ```

---

18. When Ron drained 13 quarts of water from his waterbed, it was half-filled.
    How many quarts of water could Ron's waterbed hold? (5)

19. Two hamburgers and a pizza cost $18.
    If each hamburger cost $2, find the cost of the pizza.   (5)

    $2
    Hamburger
    Hamburger $18
    Pizza

20. Sam bought a gift bag for $7.60 and a ribbon for 40¢.
    He gave the cashier a ten-dollar bill.
    How much change did he get back?   (5)

Test Yourself 3

**Do these problems.
Show your work clearly.**

1. Mr. Fu bought 8 bags of rice.
   Each bag weighed 2 pounds.
   How many pounds of rice did Mr. Fu buy? (5)

2. A pail can hold 24 quarts of water.
   4 basins of water are needed to fill the pail.
   How many quarts of water can each basin hold? (5)

3. Paul had 361 cards.
   He sold 184 of them.
   How many cards didn't he sell? (5)

4. 28 chairs are arranged in 4 equal rows.
   How many chairs are in each row? (5)

5. Mr. Carter bought 90 feet of rope.
   He used 48 feet of it.
   How many feet of rope did Mr. Carter have left? (5)

6. Wendy had $37.
   She bought a dress for $19 and saved the rest.
   How much money did she save? (5)

7. 5 girls shared 30 grapes equally.
   How many grapes did each girl receive? (5)

8. A pile of 5 identical dictionaries is 20 inches high.
   What is the thickness of each dictionary? (5)

9. When Mr. Avella poured 38 gallons of water into an empty drum, it was half-filled.
   How many more gallons of water must he pour into the drum to fill it up? (5)

10. Study the picture graph below carefully.
    Then answer the questions. (5)

Number of Grapes Each Person Ate				
Marina	△	△	△	
Thomas	△	△		
James	△	△	△	△
Dennis	△ △ △ △ △			
Brad	△	△	△	
Each △ stands for 4 grapes.				

(a) How many grapes did James eat?

(b) Who ate 20 grapes?

(c) Which two people ate an equal number of grapes?

(d) Who ate the smallest number of grapes?

(e) How many grapes did Marina and Brad eat altogether?

11. William bought a plate of pork-fried rice, a piece of cake,
    and a bowl of soup.
    He gave the cashier $50.
    How much change did he receive? (5)

Menu	
Fried rice	$ 4
Pizza	$10
Cake	$ 1
Soup	$ 3

12. My clock reads 5:40.
    It is 30 minutes fast.
    What is the actual time? (5)

13. Teddy has 47 raisins.
    Teddy has 19 raisins more than Betty.
    How many raisins does Betty have? (5)

14. Jan has $16 more than Ron.
    Jan has $36.
    How much money does Ron have? (5)

---

15. Jack has 11 grapes.
    His brother has 10 grapes.
    If they share the grapes equally with their sister, how many grapes will each of them get? (5)

---

16. Marti is 180 centimeters tall when she stands on a stool.
    The stool is 47 centimeters high.
    What is Marti's actual height? (5)

17. Morris left home at 9:30 a.m.
    He reached his cousin's house half an hour later.
    What time did Morris reach his cousin's house? (5)

18. 9 pitchers of water can fill a tank.
    3 glasses of water can fill a pitcher.
    How many glasses of water can fill the tank? (5)

19. Amos spent $41 in a sports shop.
    He spent $14 less than Pete.
    How much did Pete spend? (5)

20. Bryan wants to bicycle along the shorter road to school.
    (a) Which road is shorter?
    (b) How much shorter is it? (5)

    Road A 518 m

    Road B 945 m

Answers

Exercise 1
Level 1
1. 69
2. 20
3. 11
4. 87
5. 31
6. 33
7. 50
8. 68

Level 2
1. 27
2. 78
3. 83
4. 51
5. 76
6. 68
7. 33
8. 78
9. 51
10. 16
11. 38
12. From the diagram,
    78 − 52 = 26
    Wilson has 26 more balloons than Mellie.
    He must give half of his 26 balloons to Mellie so that both of them have the same number of balloons.
    26 ÷ 2 = 13
    He must give Mellie 13 balloons.

Exercise 2
Level 1
1. 499
2. 488
3. 227
4. 148
5. 178
6. 591
7. 174
8. 971

Level 2
1. 944
2. 210
3. 87
4. 354
5. 297
6. 270
7. 927
8. 87
9. 344
10. 682
11. From the diagram,
    2 shaded parts = 64 − 46
    = 18 beads
    Therefore, 1 shaded part = 18 ÷ 2
    = 9 beads
    Paola had 9 beads.
12. From the diagram,
    350 − 179 = 171
    Ricardo has 171 cherries left.
    171 + 246 = 417
    He has 417 cherries now.

Exercise 3
Level 1
1. 8
2. 9
3. 12
4. 28
5. 30
6. 18
7. 25
8. 40

Level 2
1. 30
2. 20
3. 35
4. 27
5. 40
6. $18
7. 15
8. 36
9. 24
10. $21
11. 70
12. Make a guess.
    Which two numbers when multiplied, give 24 and when subtracted, give 2?
    $12 \times 2 = 24$, $12 - 2 = 10$ ✗
    $8 \times 3 = 24$, $8 - 3 = 5$ ✗
    $6 \times 4 = 24$, $6 - 4 = 2$ ✓
    ♦ = 6
    ♣ = 4

Exercise 4
Level 1
1. 8, 8
2. 6, 6
3. 5, 3, 3
4. 10, 4, 4
5. 9
6. 7
7. 10
8. 8

Level 2
1. 9
2. 9
3. 2
4. 8
5. 4
6. 6
7. 6
8. 10
9. 4
10. 9
11. 5
12. 1 butterfly = 1 unit
    2 ladybugs = 1 butterfly = 1 unit
    Therefore, 4 ladybugs = 2 units
    1 butterfly + 4 ladybugs = 3 units = 30
    1 unit = 10
    1 ladybug = 10 ÷ 2 = 5

Exercise 5
Level 1
1. (b) 6
2. (a) 23 (b) 6
3. 837
4. 25
5. 78
6. 16
7. 185
8. 28

Level 2
1. 157 yards
2. 18 feet
3. 4 yards
4. 387 feet
5. 455 feet
6. 8 inches
7. 100 yards
8. 138 meters
9. 74 inches
10. 24 inches
11. From the diagram,
    264 − 178 = 86 in.
    The plank is 86 inches long.
    86 + 264 = 350 in.
    Their total length is 350 inches.
12. From the diagram,
    154 − 18 = 136 cm
    Wilson is 136 centimeters tall.
    136 + 7 = 143 cm
    Jerry is 143 centimeters.

Exercise 6
Level 1
1. 15
2. 71
3. 1,000
4. 227
5. 377
6. 923
7. 185 ounces
8. 68

Level 2
1. 364 ounces
2. 520 ounces
3. 186 pounds
4. 72 pounds
5. 576 ounces
6. 37 pounds
7. 946 grams
8. (a) 147 ounces
(b) 462 ounces
9. 6 pounds
10. 493 ounces
11. From the diagram,
    210 + 85 = 295 g
    Box B is 295 grams.
    295 + 60 = 355 g
    Box C is 355 grams.
12. 2 units = 800 − 200 = 600
    1 unit = 600 ÷ 2 = 300

    Crab | 1 unit | 200 |
    Hake Fish | 1 unit |
    } 800

    Weight of crab = 300 + 200 = 500 g

Exercise 7
Level 1
1. $5.60
2. $19.60
3. $9.90
4. $13.90
5. $24
6. 40¢
7. $11

Level 2
1. $8.80
2. DVD and pen
3. $18
4. $40
5. (a) $15.50
   (b) $25.50 − $12 = $3.50
6. $16, $8, $2, $26
7. $40.80
8. From the diagram,
   1 carton of milk = $9 − $3 = $6
   1 chicken breast = $6 ÷ 3 = $2

Exercise 8
Level 1
1. $\frac{3}{8}$
2. $\frac{1}{6}$
4. $\frac{3}{8}$
5. $\frac{1}{3}$
6. $\frac{5}{9}$
7. $\frac{5}{8}$
8. 5

Level 2
1. $\frac{2}{5}$
2. $\frac{4}{5}$
3. $\frac{5}{8}$
4. $\frac{7}{12}$
5. $\frac{3}{10}$
6. $\frac{5}{12}$
7. $\frac{3}{7}, \frac{3}{5}, \frac{3}{4}, \frac{3}{3}$
8. $\frac{1}{6}$

Exercise 9
Level 1
2. 25
3. 30
4. 55
5. 10:00 or 10 o'clock
6. 7:30
7. 2:40

Level 2
5. 30
6. 30
7. 1
8. 1
9. a.m.
10. p.m.
11. 9:30 p.m.
12. 9:30 to 10:30 is 1 hour.
    10:30 to 11:30 is 1 hour.
    1 hour + 1 hour = 2 hours

Exercise 10
Level 1
1. 33
2. 280
3. 820
4. 19
5. 332
6. 70
7. 11
8. 11

Math Challenge the Singapore Way

Level 2
1. 834 gallons
2. 349 liters
3. 80 quarts
4. 176 liters
5. 9 gallons
6. 14 liters
7. 8 pints
8. 27 liters
9. 24 gallons
10. 3
11. From the diagram,
    Pitcher B has 2 × 4 = 8 qt.
    Pitcher A has 8 × 2 = 16 qt.
12. (a) 15 – 4 – 3 = 8 L
    (b) 20 – 8 = 12 L

Exercise 11
Level 1
1. (a) Teresa (b) Mark (c) 6 (d) Mark
2. (a) Football (b) Basketball (c) 3 (d) 48

Level 2
1. (a) Friday (b) 16 (c) 20 (d) Monday
   (e) 64
2. (a) yellow (b) 10 (c) yellow (d) 40
   (e) 15 (f) 15
3. (b) 9 (c) 57
4. ✿ ✿ ✿ ✿ ✿
5. 20
6. ♥ = ☼ + ☼ + ☺
   ♥ = ☺ + ☺ + ☺ + ☺ + ☺ + ☺ + ☺
   ♥ = 7☺

Exercise 12
Level 1
1. 4
2. B, D, G
3. 6

Level 2
1. (a) 6 (b) 2 (c) 5 (d) 1

Exercise 13
Level 1
1. 4
2. (a)   (b)   (c)

Level 2
1.   2.

3. (a) semicircle (b) quarter circle (c) triangle (d) square
   (e) rectangle (f) oval

Test Yourself 1
1. 9 liters
2. $27
3. 9
4. 8
5. $1
6. 8
7. 417 grams
8. 28 feet
9. $1
10. 316 kilograms
11. 18 yards
12. $25
13. 10
14. 7 meters
15. $26
16. 10
17. 70
18. 53
19. $31
20. 182

Test Yourself 2
1. 24
2. 70
3. $54.95
4. 27 gallons
5. 8
6. $3
7. (a) 389 grams (b) 913 grams
8. 10 quarts
9. 7
10. 20
11. (a) 22 kilograms (b) 25 kilograms
12. 6
13. 181 meters
14. 286 centimeters
15. 5
16. 85 liters
17. 69
18. 26 quarts
19. From the diagram,
    2 + 2 = $4
    The two hamburgers cost $4.
    18 − 4 = $14
    The pizza cost $14.

20. $7.60 + 40¢ = 8.00
    10 − 8 = $2

Test Yourself 3
1. 16 pounds
2. 6 quarts
3. 177
4. 7
5. 42 feet
6. $18
7. 6
8. 4 inches
9. 38 gallons
10. (a) 16 (b) Dennis (c) Marina and Brad (d) Thomas (e) 24
11. $42
12. 5:10
13. 28
14. $20
15. 7
16. 133 centimeters
17. 10:00 or 10 o'clock
18. 1 pitcher holds 3 glasses of water.
    Therefore 9, pitchers will hold 9 × 3 = 27 glasses of water.
19. $55
20. (a) Road A (b) 427 meters

Notes

Notes

Notes

Notes

Notes

Notes